MW00715185

TASMANIAN LANDSCAPES

Photography by DENNIS HARDING

The mountains, forests, coastlines and heritage
that make this island unique

Dedicated in loving memory to Mollie

Published and distributed by
Harding's Productions
1 Clare Street, Hadspen,
Tasmania 7290 Australia
Phone: 03 6393 6955
Fax: 03 6393 7288

ISBN 0 9579254 2 5

Graphic design and illustrations by Jennifer Morrison, Jaffa Design

Printed by Everbest Printing Co Ltd, China

I would sincerely like to thank the following people who have helped me along the way;
Rob Blakers, Barry Ashenhirst, Doug Simpson, Paul Burger, Greg Wood, Ken Duncan,
Lionel Marz, Peter Brown, Jennifer Morrison, Jean-Paul Ferrero, Vere Kenny, Leslie Richards,
Kevin House, John Crook, Greg Willson, Mark Beaton, Frits deBruyn, John Temple,
Chris Viney, Wayne Morgan and Rolfe de la Motte.
Special thanks to my wife Barbara and my son Michael for their consistent ongoing support.

www.dennisharding.com.au

Front cover: Deciduous beech (Nothofagus gunnii) beneath Mt Olympus
Inside front cover: Mt Murchison, West Coast
Title page: Old-growth forest, Southwest Tasmania
Contents page: Mount Kate Hut, Cradle Valley
Back cover photos clockwise from top left: Rainforest, Lady Barron Creek, Mt Field National Park;
Sunrise at Cradle Mountain; Early morning light on the Penitentiary, Port Arthur;
North of Bicheno, East Coast

CONTENTS

TOWNS & GARDENS

THERE ARE SURPRISES CLOSE TO THE HEART OF THINGS IN TASMANIA'S FRIENDLY, PEOPLE-SIZED CITIES AND QUIET, TREE-SHADED COUNTRY TOWNS.

A FEW MINUTES FROM LAUNCESTON'S CITY CENTRE, THE RAPIDS OF A WILD RIVER RUSH THROUGH A BOULDER-CHOKED GORGE - WHILE A SHORT DRIVE FROM THE BUSTLE AND COLOUR OF HOBART'S SALAMANCA MARKET ARE THE CRAGGY ALPINE LANDSCAPES OF A LOFTY MOUNTAIN, CAPPED WITH A WHITE MANTLE OF SNOW THROUGH THE WINTER.

IN THE GENTLE COUNTRYSIDE, SOLID BUILDINGS OF HAND-CUT FREESTONE STAND FIRM AND FOUR-SQUARE, LONG AFTER THE LAST ECHOING CLINK OF CONVICT CHISELS HAS FADED. SLOW-FLOWING STREAMS SWIRL BENEATH SANDSTONE BRIDGES AND THE SPREADING BOUGHS OF ENGLISH TREES LAY DAPPLED LIGHT ON CALM WATER.

AND IN TASMANIA'S PARKS AND GARDENS, EVERY SEASON HAS ITS JOYS - A SCATTER OF AUTUMN LEAVES, RED-GOLD ON GREEN LAWNS; A BLAZE OF WINTER WATTLE; THE FRESH GREEN BURST OF SPRINGTIME BUD AND BLOSSOM; ALL THE BRIGHT BLOOMS OF SUMMER FLOWERS.

Mt Wellington from Lindisfarne Bay

SALTAIR
ROSEVEARS TAVERN

Hobart and the River Derwent from the Pinnacle, Mt Wellington

Early morning light on the Penitentiary, Port Arthur

City Park, Launceston

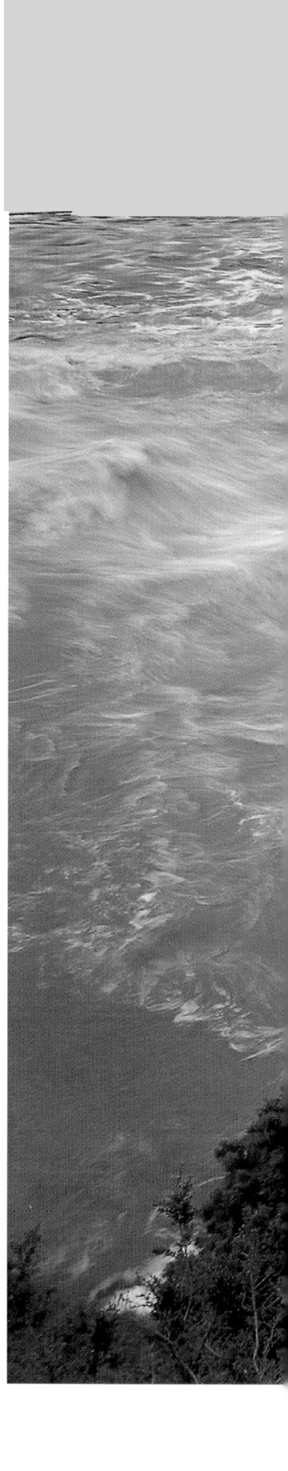

Floodwaters in Launceston's Cataract Gorge

The Conservatory, Royal Tasmanian Botanical Gardens, Hobart

Franklin Square, Hobart

Allendale Gardens, Edith Creek

South Esk River, Perth

St Andrews Church, Evandale

Vineyard in the Coal Valley, Richmond

Bridge Street, Richmond

The Salmon Ponds at Plenty, Derwent Valley

Convict carvings on the Ross Bridge

Kings Bridge, Launceston

Brickfields Reserve, Launceston

The Nut, Stanley

ROADSIDE SCENES

IN ONE DAY'S EASY DRIVING, TRAVELLERS CAN FOLLOW TASMANIAN ROADS FROM SEA TO SEA, PASSING THROUGH FARMLANDS AND FORESTS, MOUNTAINS AND WORLD HERITAGE WILDERNESS, TO REACH ANOTHER COAST ON THE OTHER SIDE OF THE ISLAND.

INSTEAD OF FLAT AND EMPTY HORIZONS, TASMANIA'S ROADSIDE LANDSCAPES ARE RICHLY VARIED - AT EVERY CREST AND TURN, NEW SCENERY EMERGES.

THIS IS A DESTINATION MADE FOR LEISURELY TOURING - WHILE THE AIM OF THE DAY MAY BE TO LEAVE ONE PLACE AND REACH ANOTHER, THE VIEWS TO BE ENJOYED DURING THE DRIVE ARE ALWAYS AS MEMORABLE AS THE ARRIVAL.

BEYOND THE ROADSIDES, HILLS MERGE INTO DISTANT MOUNTAIN RANGES; VALLEY ROADS FOLLOW THE SINUOUS CURVES OF TREE-LINED RIVERS; ROUTES INTO THE HIGHLANDS SKIRT THE BANKS OF ALPINE LAKES; COAST ROADS HUG THE SEA, WITH GLIMPSES TO ISLANDS FLOATING ON THE HORIZON; COUNTRY BYWAYS WANDER ALONGSIDE PADDOCKS STRIPED WITH FLOWERS; WHEN CANOLA CROPS ARE IN BLOOM, HIGHWAYS RUN THROUGH FIELDS OF GOLD.

Fog at Beam's Hollow, near Hadspen

Late afternoon light on farmland near Weetah

Canola field near Conara

Convict-built ruins, Oatlands

Poppy field, Table Cape

Sunrise on Mt Roland

Huon River rainbow

Flowering peas, Sassafras

Bridge on Picton River

Hollybank Forest near Lilydale

Abandoned farmhouse, NW Tasmania

Tulip farm, Table Cape

Central Highlands tarn

Autumn on the River Derwent, Bushy Park

North Esk River at Corra Linn

COASTLINES

THIS HEART-SHAPED ISLAND IS CARVED AND DEFINED BY THE SEA - THE RESTLESS WAVES OF BASS STRAIT, THE AZURE WATERS OF THE TASMAN, THE ENDLESS GREEN GROUNDSWELLS OF THE VAST SOUTHERN OCEAN.

WAVES THAT HAVE CIRCLED THE EARTH'S HIGH SOUTHERN LATITUDES CAST THEIR ENERGY ON THE BEACHES AND CLIFFS OF THE ISLAND'S SOUTHERN AND WESTERN COASTLINES. TO THE EAST, SQUEAKING WHITE SAND AND LICHEN-SPLASHED RED GRANITE MEET THE SEA.

ON THE MARGIN BETWEEN LAND AND WATER, THE INTERACTION OF WAVE AND TIDE ON ROCK AND SAND CREATES A RICH VARIETY OF LANDSCAPES - THE TASMAN PENINSULA'S BEETLING GREY SEA CRAGS, SHEER FROM SEA BED TO SUMMIT; WAVE-TUMBLED COBBLES ON THE RIM OF BLUESTONE BAY; THE GLEAMING WHITE QUARTZITE CLIFFS AND BEACHES OF THE SOUTH COAST; THE SOFT FORMS OF HARD GRANITE ON THE HAZARDS; LONG STRETCHES OF SAND, CRUSHED FROM PARENT ROCK BY THE RELENTLESS SWISH OF WATER THROUGH UNCOUNTABLE YEARS.

AND ON EVERY TASMANIAN SHORELINE, THE SOUND OF THE SEA FILLS THE SENSES - ROARING, WHISPERING, SOBBING, SIGHING.

North of Bicheno, East Coast

Sunrise, Ketchem Bay,
South Coast

Near Granville Harbour, West Coast

Morning at Friendly Beaches, Freycinet National Park

Early light on sea cliffs,
Cape Hauy

Bluestone Bay, Freycinet Peninsula

Eddystone Point, East Coast

Stream at Ketchem Bay, South Coast

Hidden Bay, South Coast

The Hazards from Coles Bay, Freycinet Peninsula

t Cameron West,
est Coast

Sunrise on Freycinet, Cape Tourville, East Coast

The Hazards from Richardsons Beach, Freycinet Peninsula

The Blowhole at Bicheno, East Coast

Lion Rock, South Cape Bay, South Coast

Afterglow on Picnic Rocks, Mt William National Park, East Coast

FORESTS

UNDER A PALE ANCIENT SUN, SWATHES OF GREAT TREES FLOURISHED, GREW OLD, DIED OR BURNED, GREW OLD AGAIN. AND AGAIN.

IN TASMANIA'S WET EUCALYPT FORESTS, THE TALLEST FLOWERS ON THE PLANET INCH INTO THE SKY - THREE HUNDRED YEARS FROM SEEDLING TO STAG. IN THE WESTERN RAINFORESTS ARE THE GNARLED TRUNKS OF SOME OF THE EARTH'S OLDEST LIVING THINGS - TREES THAT COULD HAVE BEEN PUSHING THROUGH THE LITTER WHEN TUTANKHAMEN SAT ON THE SUN KING'S THRONE, MORE THAN 3000 YEARS AGO.

BENEATH ALL THE LAYERS OF THE YEARS - BARK AND TWIG, MOSS AND PEAT, CINDERS AND CHARCOAL - A MYRIAD OF BUSY BEINGS LIVE OUT THEIR SHORT AND VIOLENT LITTLE LIVES IN THE SOIL, TURNING LEAF-FALL INTO NUTRIENTS FOR THE NEXT GENERATION OF FOREST GIANTS.

IN THE GREEN AND DRIPPING GLOOM OF THE FORESTS, TIME SWALLOWS SOUND - EXCEPT IN PLACES WHERE SHARP-VOICED MACHINES MOVE UP THE ROAD AND INTO THE TREES.

Little Fisher Valley

Ancient myrtle, North East Tasmania

*Forest cascade,
Meander Valley*

Waterfall, Liffey Valley

Fungi, Notley Gorge

Tree ferns, Meander Valley

Giant Eucaly
Great Western Ti

Waratah seedling, Meander Valley

Snow on tree fern, Split Rock Waterfall, Great Western Tiers

Swamp gum (Eucalyptus regnans), Styx Valley

Rainforest, Lady Barron Creek, Mt Field National Park

Cascade, Liffey River

Eucalypt forest at sunset near Ragged Jack

Ancient Huon pine, Huon River

New Falls Creek, Southwest Tasmania

Myrtles in Meander Valley

Snow covered myrtle, Meander Valley

HIGH PLACES

BEYOND THE SERENITY OF TASMANIA'S MOUNTAINS ARE THE EARTH-SHATTERING EVENTS THAT BUILT AND SHAPED THEM. THE ISLAND'S COMPLEX GEOLOGY IS A TEXT BOOK WITH OPEN PAGES.

TURN BACK TO THE EARLIEST PRE-CAMBRIAN TIMES, WHEN SANDY SEDIMENTS SETTLED QUIETLY ON THE BED OF A LIFELESS SEA. CRUSHED, HEATED, FOLDED AND RAISED, THEY STAND TODAY AS THE GREAT JAGGED TEETH OF FEDERATION PEAK, FRENCHMANS CAP AND ALL THE WHITE QUARTZITE MOUNTAINS OF THE SOUTHWEST.

LEAF AHEAD A FEW MILLION PAGES TO THE JURASSIC, WHEN DINOSAURS RULED THE EARTH. AS THE SOUTHERN SUPERCONTINENT GONDWANA SPLIT APART, A MASSIVE UPWELLING OF MOLTEN ROCK SURGED INTO THE CRUST. HARDENING BENEATH OLDER ROCKS AND EXPOSED ON THE SURFACE AFTER MILLIONS OF YEARS OF EROSION, THE DOLERITE FORMS SHEER GREY CLIFFS FROM MT WELLINGTON TO CRADLE MOUNTAIN.

THEN ICE WRENCHED THE ROCKS, CARVED THE ESCARPMENTS, SCOOPED THE BED OF LAKES, CARRIED GREAT BOULDERS MILES FROM WHERE THEY TUMBLED.

TODAY, THE GLACIERS HAVE RETREATED, BUT THE FORCES OF WATER AND ICE STILL SHAPE THE HIGH COUNTRY, AS FROST FORMS IN THE CRACKS OF CRAGS AND CREEKS GURGLE UNDER THE SNOW - WASHING, WEDGING, WEARING, CARRYING THE MOUNTAINS AWAY, GRAIN BY GRAIN.

Alpine tarn near Mt Gould

Mt Murchison, West Coast

Waterfall Valley, Cradle Mountain-Lake St Clair National Park

Mt Oakleigh from Pelion Plains

Sunset on the crags of Mt Geryon from the Labyrinth

Pencil pine,
Central Plateau

West Wall from the Pool of Siloam, Walls of Jerusalem

Forest beneath Mt Gould

Waterfall in pre-dawn light, Cradle Plateau

Lake Oenone, Cradle Mountain-Lake St Clair National Park

Sunrise at
Cradle Mountain

Pinnacle Ridge, Mt Oakleigh

Deciduous beech (Nothofagus gunnii) beneath Mt Olympus

Twisted Lakes,
Cradle Mountain

Wombat Tarn, Cradle Mountain

Federation Peak, Eastern Arthur Range

Meander Falls

Cradle Mountain from Dove Lake